Introduction to polymer dynamics

Lezioni Lincee
Editor: Luigi A. Radicati di Brozolo, Scuola Normale Superiore, Pisa

This series of books arises from a series of lectures given under the auspices of the Accademia Nazionale dei Lincei through a grant from IBM.

The lectures, given by international authorities, will range on scientific topics from mathematics and physics through to biology and economics. The books are intended for a broad audience of graduate students and faculty members, and are meant to provide a '*mise au point*' for the subject they deal with.

The symbol of the Accademia, the lynx, is noted for its sharp sightedness; the volumes in the series will be penetrating studies of scientific topics of contemporary interest.

Introduction to polymer dynamics

PIERRE GILLES DE GENNES

Collège de France, Paris

CAMBRIDGE
UNIVERSITY PRESS

Published by the Press Syndicate of the University of Cambridge
The Pitt Building, Trumpington Street, Cambridge CB2 1RP
40 West 20th Street, New York, NY 10011-4211 USA
10 Stamford Road, Oakleigh, Melbourne 3166, Australia

First published 1990
Reprinted 1992, 1995

Printed in Great Britain by Athenæum Press Ltd,
Gateshead, Tyne & Wear

British Library cataloguing in publication data

Gennes, Pierre Gilles de
Introduction to polymer dynamics.
1. Polymers. Dynamics
I. Title
547.7′045413

Library of Congress cataloguing in pubication data

Gennes, Pierre Gilles de
Introduction to polymer dynamics / Pierre Gilles de Gennes.
    p.   cm. — (Lezioni lincee)
Includes bibliographical references.
ISBN 0 521 38172 X. - ISBN 0 521 38849 X (paperback)
1. Polymers. I. Title II. Series.
QD381.8.G46 1990
547.7—dc20   89-37312   CIP

ISBN 0 521 38172 X hardback
ISBN 0 521 38849 X paperback

Contents

Foreword

The Lezioni Lincee on Topics of Polymer Dynamics were delivered by Professor Pierre Gilles de Gennes at the Politecnico di Milano in December 1986 to an audience of faculty members and graduate students.

The subject of polymer dynamics has been expanding at a very rapid pace in the last few decades and Professor de Gennes has been an active contributor to the field. In particular, several new theoretical ideas have appeared, one of their purposes being to connect different phenomena into a unified frame.

Rather than attempting to give a complete discussion of the whole subject, Professor de Gennes has concentrated on some special topics.

Chapter 1 (which consists of the first and the second lecture as originally delivered in Milan) contains a discussion of some general aspects of polymer chain dynamics. After a brief general illustration of the subject, the motion of isolated chains in a dilute solution is analyzed. Global and local deformation modes are discussed, with reference both to uncharged chains and to flexible polyelectrolytes. Attention is then focused on polymer melts; their viscoelastic behavior is seen in the light of the concept of chain 'reptation'. Chemical kinetics in entangled media is then taken into consideration, with several examples. Some other aspects of polymer dynamics are also analyzed, including crystallization and the behavior of ramified chains.

The subject of Chapter 2 (originally the fourth lecture) is the conformation of a protein chain around a receptor site, a problem of great importance in molecular biophysics. Professor de Gennes uses a statistical approach to derive some relevant conformational parameters of the problem, most important the minimum number of aminoacids comprised in one loop. This protein study was performed more than fifteen years ago, but appeared only as an oral remark in a biology conference.

In Chapter 3 entitled 'Dry spreading of liquids on solids' (originally the last lecture) the general phenomenon of surface wetting by liquids with different spreading coefficients is examined in its most basic aspects. The specific case of polymer melts is also taken into consideration: here we are confronted with many new possibilities: either an anomalous *slippage* of the melt at the wall surface or in some other cases an anomalous *sticking* of the chains to the wall.

The subject of Chapter 4 (originally the third lecture) is the problem of the reduction of turbulent losses by polymer chains in a solvent. The topic appears to be especially intricate in view of the coupling between the hydrodynamic aspects of turbulent flow and the viscoelastic behavior of chains in strongly perturbed conformations. The lecture followed closely a paper 'Towards a scaling theory of drag reduction' published in 1986 by Professor de Gennes in *Physica* which is here reprinted by permission of the publisher. Because this lecture is the most tentative (and difficult) part in the series we have put it in the last chapter.

In many of these problems, nature may have more than one answer to our queries, depending on the system chosen. But these discussions may help us to list some of the possible answers, and to raise new questions.

GIUSEPPE ALLEGRA

1

General aspects of polymer chains

1.1 Physical states of molecules in long chains

We begin with some specific examples.

(a) Polyethylene:

$$\ldots -CH_2-CH_2-CH_2- \ldots, \text{ or } [CH_2]_N$$

The chemical sequence is *linear* (i.e., without branching) and built up from N monomeric units.

(b) Polystyrene:

$$[-CH_2-CH-]_N$$

(c) Polyvinylchloride:

$$[-CH_2-CH-]_N$$
$$\underset{Cl}{|}$$

These three materials are present in everyday life. Their production exceeds that of metals, although, unlike metals, they have been known for only half a century! We notice from the outset that the number N of units in a molecule can be very large (up to 10^5). From a chemical point of view, the construction of a chain with this kind of length implies repeating (up to 100 000 times) the same elementary operation *without mistakes*: a remarkable *tour de force*. Common polymers do have mistakes: they have many defects along the chain! It is, however, possible to avoid the presence of defects at the level of fundamental research, using very elegant chemical techniques (see, for instance Champetier and Monnerie, 1969).

The first steps in chemical research on polymers have been quite difficult. The synthesis of macromolecules was established by the patient work of Staudinger. However, Staudinger's work met with strong resistance before being accepted. Staudinger himself was mistaken on one point: he was convinced that all linear macromolecules form rigid sticks. Somewhat later, Kuhn had the merit of discovering that the majority of linear polymers in solution (or in melts) are actually quite flexible: this flexibility gives them a large variety of properties.

What are the physical states of a flexible linear polymer?

(a)   At high temperature, it forms a *liquid* of entangled chains.

(b)   Upon cooling, one obtains a *glass* most of the time: it is precisely this glass that we utilize for instance with the polystyrene of some plastic packing.

(c)   Under some favorable circumstances, the polymer may *crystallize* upon cooling. The mechanism of this crystallization is quite complex, and is still the object of intense discussion (*Faraday Transactions*, 1979). Often, if one starts from the liquid phase, the crystallization is only partial. However, this leads to interesting mechanical properties: 'spontaneous composites' are formed with crystalline parts in a glassy matrix.

The 'metallurgy of polymers' is currently developing, and it will become as rich as the metallurgy of metals, with some interesting differences. One of these differences concerns the *alloys*: it is tempting to mix two polymers in order to improve their properties. However, obtaining these alloys turns out to be quite difficult.

The number of polymer pairs (A, B) that can be mixed in the liquid phase is very small. This is due to Van der Waals forces: see, for instance, de Gennes (1984). Fortunately, there exist alternative methods to hybridize the properties of A and B. The most common consists of producing some 'statistical copolymers'

$$\dots ABBABAAA \dots$$

These copolymers are essential, for instance, in the manufacture of synthetic rubbers. A different approach (more expensive, albeit interesting for the future) makes use of 'block copolymers':

$$[AA \dots A]_N \text{-} [BB \dots B]_M.$$

A double sequence A–B of this kind arranges itself spontaneously at the interphase between liquid phases A and B: it lowers considerably the surface tension, and it enables one to obtain interesting *emulsions* of A in B (or vice-versa).

Another very rich aspect of the metallurgy of polymers is given by the effects of *vulcanization*: starting from a liquid of linear chains, one joins together some chains that happen to be close to each other, with a suitable chemical operation. One obtains a random lattice of connected chains which, locally, is still a fluid, but which, at a macroscopic level, resists compression with a non-zero elastic modulus: a *rubber*.

The above examples illustrate the variety of physical states which can be realized with flexible chains, the many applications to which they can lead, and also the fundamental problems of statistical mechanics that they raise. In the following, we shall face first the problem of *dilute* chains in a solvent, which is important in order to *characterize* polymers: to determine the molecular weight (i.e. the length of a chain) one often makes use of dilute solutions. We shall then proceed to the melts, the understanding of which is essential for many practical purposes.

1.2   Isolated chains

1.2.1   *Overall motion of an isolated chain*

Let us consider first a single coil which is immersed in a solvent.

In 1948 Debye and Bueche proposed to describe this from a hydrodynamical standpoint, in terms of a porous sponge with uniform density. They found that vorticity cannot penetrate this sponge (except for a certain 'screening length' which is negligible for a long chain). They concluded that, as far as global properties (such as viscosity and sedimentation) are concerned, the coil possesses a hydrodynamic radius R_h which is proportional to its radius of gyration.

This result is qualitatively correct in three dimensions; it was confirmed in a more sophisticated analysis by Kirkwood. However, the same sponge model is incorrect for other situations. F. Brochard considered the case of one coil confined within a very narrow tube and showed that the Debye approach is qualitatively wrong in this case. The total mobility of the chain does not have the structure and the dependence on molecular weight that one would expect from the sponge model: the actual mobility is much higher. Physically, this occurs because of local concentration fluctuations in the interior of the coil: at any instant, there exist in the coil some 'channels' with few monomers, allowing for easier permeation of the solvent.

1.2.2   Global relaxation times of a chain

By means of mechanical, mechano-optical, or (under certain favorable circumstances) dielectric, measurements, we can assign to a chain (in dilute solution) a global relaxation time τ_1. A suggestive description of τ_1 makes use of the dumbbell model of Kuhn and Péterlin, according to which one describes the chain by its elongation $\vec{r}$ only. For small values of $\vec{r}$, Péterlin wrote down a force balance of the form

$$f\frac{d\vec{r}}{dt} + K\vec{r} = 0 \tag{1}$$

where f is a global friction coefficient and K is an elastic constant. This gives:

$$\tau_1 = f/K \tag{2}$$

The problem is to assign a functional form to f and K, and specifically to predict the dependence on the molecular weight. (a) The first attempt (Kuhn–Rouse) assumed friction to be additive $f = Nf_m$ (where N is the number of monomers per chain) and the elastic forces to coincide with those of an ideal gaussian chain

$$K \cong \frac{k_B T}{R_0^2} = \frac{k_B T}{Na^2} \quad \text{(ideal chain).} \tag{3}$$

There follows a relaxation time which is proportional to the square of the molecular weight

$$\tau_1 \cong N^2 \frac{a^2 f_m}{k_B T} \tag{4}$$

(where T is the temperature and k_B is the Boltzmann constant). (b) In fact, friction is not additive: using Debye's language, if we consider a coil in Stokes flow the monomers in the 'interior' are screened from the flow, and only the monomers in the 'exterior' are subject to friction. We find the same property for the global deformation mode. The friction coefficient therefore takes the form

$$f \cong \eta R \tag{5}$$

where R is the coil size and η is the solvent viscosity. Let us consider an ideal chain ($R = R_0$): we then arrive at the Zimm expression

$$\tau_1 \cong \frac{\eta R_0^3}{k_B T} \cong \frac{\eta a^3}{k_B T} N^{3/2} \quad \text{(ideal chain).} \tag{6}$$

(c) How would we extend these results to the more common case of a chain in a good solvent? In this case, owing to excluded volume effects, we expect

$$R \cong R_F = N^{3/5} a, \tag{7}$$

where R_F is what we call the Flory radius. This value of R is to be inserted in the friction coefficient f. However, and this is the essential point, when we deal with a good solvent, *we must also correct the elastic constant K.* Taking into account the scaling laws for a swollen coil (see de Gennes, 1977), we obtain the free energy F as a function of the elongation r in the form

$$F = k_B T g\left(\frac{r}{R_F}\right),$$ (8)

where $g(x)$ is a dimensionless function. For small x, $g(x)$ has the form $g_0 + g_2 x^2$ (Hooke's law), from which it follows that the elastic constant is given by

$$K \cong \frac{k_B T}{R_F^2} \cong \frac{k_B T}{N^{6/5} a^2} \quad \text{(good solvent)}.$$ (9)

We obtain a time

$$\tau_1 \cong \frac{\eta R_F^3}{k_B T} \quad \text{(good solvent)}.$$ (10)

We can relate the coil volume R_F^3 to the viscosity in dilute solutions: the viscosity increase $\delta\eta$ due to a small concentration c in polymers has the form

$$\delta\eta \cong \eta \frac{c}{M} R_F^3.$$ (11)

Introducing then $\delta\eta/\eta c = [\eta]$, we recast τ_1 in the form

$$\tau_1 \sim [\eta] M / k_B T.$$ (12)

Equation (12) is widely verified in the literature. However, the community of scientists working on polymers has not always realized that its validity in a good solvent relies on the scaling law (8) and thus on a non-trivial form of the elastic energy (9).

1.2.3   Internal deformation modes

We consider again an isolated coil in a good solvent, and focus our attention on frequencies much larger than $1/\tau_1$, or spatial dimensions much smaller than the coil radius R_F.

One can attempt to define internal *modes*, by writing a dynamical equation that generalizes equation (1) for each monomer, including the following main features.

(a) 'Back flow': each monomer, by its displacement, generates in the solvent a long-range velocity field which influences the other monomers; the effect has been calculated by Zimm.

(b) Renormalization of elastic forces: akin to passing from equations (3) to (9) for the dumbbell force constant; this effect has been included in de Gennes (1976–77).

However, the modes just defined do not have a very precise meaning. According to classical mechanics, the notion of modes requires the relaxation time spectrum of a chain (with given N) to consist of a discrete series of peaks. Actually, owing to the non-linear coupling among modes that is introduced by the above effects (a) and (b), the spectrum is a continuum, and one cannot expect sharp peaks: *the notion of modes is probably meaningless.*

Fortunately, the dynamical scaling laws still enable us to make predictions. The pioneering experiments in this field are due to a collaboration between research groups in Strasbourg and in Saclay.

They made use of very long molecules of polystyrene ($M = 3 \times 10^7$) the radius, R_F, of which can be made large in comparison with the optical wavelength. More precisely, typical values of the wave vector q lead to values $qR_F \sim 10$ in light-scattering experiments: under these conditions one can probe the internal modes. In an inelastic scattering experiment, with given q, the spectrum of the outgoing light $S(q, \omega)$ (where ω is the frequency transfer) turns out not to be Lorentzian. There is a well-defined half width Γ_q (that increases rapidly with q). The experimental results are in rather good agreement with an old calculation carried out for ideal chains, yielding $\Gamma_q \sim q^3$. Even the detailed structure in ω of $S(q, \omega)$ seemed to be correct! This is rather surprising at first glance because we are dealing with a good solvent; the chains are not ideal. Scaling laws can help us understand the generality of the result. They are obtained in the following way (de Gennes, 1977):

(a)   when qR_F is small, the width Γ_q behaves as a diffusion width $\Gamma_q = Dq^2$. We shall assume that the diffusion coefficient D is of the Stokes–Einstein form

$$D = \text{const.} \frac{k_B T}{6\pi\eta R_F}; \tag{13}$$

(b)   for arbitrary qR_F, we postulate the form

$$\Gamma_q = Dq^2 \gamma(qR_F), \tag{14}$$

where $\gamma(x) \to 1$ when $x \to 0$ and $\gamma(x) \to x^m$ when $x \to \infty$.

The exponent m is then determined by the following requirement: when g is large, one probes the dynamics of small regions (with size g^{-1}) in the interior of the ball. These dynamics should be totally independent of the total length of the chain (i.e., of N or of $R_F = aN^{3/5}$). If we require equation (14) to become independent of R_F for large q, we must choose an exponent $m = 1$:

$$\Gamma_q \cong \frac{k_B T}{\eta} q^3. \tag{15}$$

This structure is quite well confirmed by the results of Adam and Delsanti.

An additional subtle point arises at the theta point Θ. For the static properties of the chains, the point Θ defines the simplest situation: the chains are almost ideal. The dynamics at $T = \Theta$, however, are rather tricky. In a good solvent, a coil is very swollen, and one can show that practically no knot occurs on it. In a Θ solvent, on the other hand, a coil is much more compact: it has a number of knots of order $N^{1/2}$, i.e., quite large. Under these conditions, the dynamics at the Θ-point can be severely modified by entanglement effects. The Zimm description, which historically appeared to be rigorous at the Θ-point, is probably not quite correct.

1.2.4 The particular case of flexible polyelectrolytes

Polyelectrolytes are chains which carry a certain amount of charge. In the presence of salt (i.e., when the screening radius is small in comparison with the chain size) polyelectrolytes arrange themselves in almost classical coils. We could then expect that the friction coefficient f is always given by a Stokes-type expression. If we apply a field E, a polyelectrolyte with charge Q should move with velocity

$$V = QE/f, \tag{16}$$

from which (since $Q \sim N$) the mobility V/E would be proportional to $N^{1-3/5} = N^{2/5}$ because of equation (7). But this is not correct. The Stokes equation (5) assumes that back-flow effects (namely, the hydrodynamic action of one monomer upon another at a distance r) have a slow $(1/r)$ spatial decay: here, however, the counter ions (that are pushed in an opposite direction to Q) screen the back-flow currents beyond the screening length and *one recovers the Rouse regime*. A similar phenomenon occurs in the Soret effect of neutral macromolecules.

1.3 Molten linear polymers: mechanical and physical properties

1.3.1 Viscoelasticity

Molten polymers flow like (highly viscous) ordinary liquids when they are acted upon by very slow perturbations. However, at slightly higher frequencies ω, they behave like a rubber. The threshold frequency $\omega_l = 1/\tau$ that separates the two spectral regions is strongly dependent on the chain length

$$\tau = \tau_0 N^a, \tag{17}$$

where τ_0 is a microscopic time ($\sim 10^{-10}$ s in liquid phase) and a is an empirical exponent ranging between 3 and 3.4. Owing to the factor N^a, τ can become quite large (milliseconds to seconds).

1.3.2   The reptation concept

The viscoelastic behavior described above has been qualitatively understood for a long time. At small times $t < \tau$, the knots within the chains to not have time to break, resulting in a rubbery behavior. At large times $t > \tau$, the knots open by Brownian motion: the chains can slide past one another, resulting in a behavior similar to a liquid. A coherent theoretical estimate of the exponent a has taken a long time. The 'reptation' model (de Gennes, 1971; 1984) yields $a = 3$ by a very simple argument: every chain, at a given instant, is confined within a 'tube' as it cannot intersect the neighboring chains. The chain thus moves inside the tube like a snake. The motion in the tube is characterized by a *mobility* μ_{tube} inversely proportional to the length of the chain, $\mu_{tube} = \mu_0 N^{-1}$. (To pull a wet string inside a tube at a given velocity, we need to apply a force proportional to the length of the string.) We can associate to the mobility μ_{tube} a diffusion coefficient (D_{tube}) that describes the Brownian motion in the tube. The displacement s at time t obeys the law:

$$\langle s^2 \rangle = 2D_{tube} \tag{18}$$

A famous equation by Einstein relates diffusion and mobility:

$$D_{tube} = kT\mu_{tube}, \tag{19}$$

T being the temperature. Having established these concepts, we can at this point define τ as the time necessary for the chain to move across the length L of the initial tube:

$$D_{tube}\tau \cong L^2, \quad \tau \cong \frac{L^2}{D_{tube}}. \tag{20}$$

Since $L \sim N$ and $D_{tube} \sim N^{-1}$, we obtain $\tau \sim N^3$. Experimentally one often finds an exponent slightly larger than 3: this disagreement has not yet been fully understood.

The reptation concept has subsequently led to the formulation of a detailed theory of viscosity, due to Doi and Edwards, which seems to explain a sizable part of the experimental data, not only for small deformations but even for non-linear phenomena. Today, the liquid linear polymers are becoming model systems for rheology.

1.3.3 Self-diffusion and adhesion

Let's assume that we have been able to 'tag' a particular chain in the solution: either by an isotopic replacement (deuterium/hydrogen), or by attaching a specific dye. We can follow the motion of the tagged chain over a large scale: it will move by Brownian motion, with a displacement $\langle x^2 \rangle$ that increases linearly with time:

$$\langle x^2 \rangle = 2Dt \quad (t \gg \tau). \tag{21}$$

The coefficient D is known as a self-diffusion coefficient; its value is much smaller than for ordinary molecules: the chains can reptate through their neighbors only rather slowly. We stress the difference between equations (21) and (18). In equation (18) we describe the *curvilinear* displacement along a tube, while in equation (21) we describe displacement at large scales along a straight line. The coefficient D is thus much smaller than D_{tube}. In fact, when $t = \tau$ we should have $x \sim R_0$, where R_0 is the gyration radius of the tagged chain. Since this chain forms an ideal coil, we have $R_0 \sim N^{1/2}$ (see, e.g., Flory (1971), de Gennes (1984) for a discussion of R_0). This gives:

$$D = \frac{R_0^2}{\tau} \sim N^{-2}. \tag{22}$$

These diffusion coefficients were first measured by J. Klein through isotopic tagging and detection by infrared spectroscopy; later on, they were measured in a more refined way by Hernet, Léger and Rondelez through photochromic tagging and the forced Rayleigh effect. Now the N^{-2} law seems confirmed by many experiments.

A different type of experiment was performed by Laurence *et al.* with a pair of compatible polymers $A + B$ (where A is polyvinyl chloride and B is polycaprolactone). To their surprise, they found $D \sim N^{-1}$ rather than $D \sim N^{-2}$. In fact, their results are quite normal (de Gennes, 1982): the mixture $A + B$ is not ideal (in contrast with the experiments discussed previously). A molecule of type A prefers to be in contact with molecules of type B. Thus, a region which is rich in A can be more readily reached by B molecules than in an ideal mixture. When we include these thermodynamic effects, we get exactly $D \sim N^{-1}$. Similar changes are possible when studying the *adhesion* of two polymers A and B (de Gennes, 1982).

1.4   Chemical kinetics in entangled media

1.4.1   A few examples

Let us imagine a reagent X to be attached to a chain, and a different reagent Y to be attached to another chain (which is chemically identical to the first one). We wish to study the kinetics of a reaction of the type:

$$X + Y \rightarrow XY, \tag{α}$$

or the deactivation of an excited species X into the species Y:

$$X^{*} + Y \rightarrow W + Y. \tag{β}$$

The first case is in practice quite common.

(a) Many syntheses of polymers involve free radicals, and are limited by the recombination of the two radicals X and Y ('termination' reactions);

(b) an irradiated polymer develops free radicals, which in turn can react among themselves and form bridges between chains.

Deactivations $(X^* + Y)$ are particularly interesting at a fundamental level, since the final product $(X + Y)$ is still formed by two independent chains: the medium is not altered by the reaction.

In the following, we shall limit ourselves for the sake of simplicity to optical deactivation: reagents X and Y are attached to two flexible chains and immersed in a liquid of the same chains. We shall also assume that the reaction occurs whenever the distance between the two centers X and Y becomes smaller than a given *trapping radius b*.

1.4.2 The small molecules case

When X and Y are not attached to any chain, but are small molecules (with concentrations n_x and n_y) in a dilute solution, the kinetics is simple:

$$\frac{dn_X}{dt} = \frac{dn_Y}{dt} = -kn_Xn_Y, \tag{23}$$

with

$$k = 4\pi b(D_X + D_Y), \tag{24}$$

where D_X and D_Y are the diffusion coefficients of the two species X and Y.

How could we transfer these results to our entangled molecules? It is initially tempting simply to introduce into equation (24) the macroscopic diffusion coefficients defined by (21) and (22). The constant k should then be proportional to N^{-2}. We can, however, show that this idea is wrong. The reaction occurs at spatial distances between reagents of the order of b, which are smaller than the statistical size R_0 of the chains; an analysis solely based on macroscopic concepts is not sufficient.

1.4.3   *Transport by reptation*

The complete discussion of migration of the reagents through 'tubes' is quite complex. We shall here give a sketchy, albeit representative, overview based on the notion of a diffusion coefficient $D(r)$ *that depends on the spatial coordinates.* We choose the variable r to be the distance between two reagents. The flux at a distance r is given by:

$$4\pi r^2 \left(-D(r) \frac{\partial c}{\partial r} \right) = J. \tag{25}$$

In a stationary regime (at large times, $t > \tau$) this flux is constant: we can then obtain the shape $c(r)$ of the concentration. The condition to be applied to the reaction is

$$c(b) = 0. \tag{26}$$

On the other hand, $c(\infty) = n_X n_Y$.

From equations (25) and (26) we can arrive at a generalized version of equation (24):

$$\frac{4\pi n_X n_Y}{J} \frac{4\pi}{k} = \int_b^\infty \frac{dr}{r^2 D(r)} \quad (t > \tau). \tag{27}$$

The coefficient $D(r)$ is the sum of the two coefficients due to the chains that carry the two reagents. Here we assume the chains to be identical and we make the replacement $D(r) \to 2D(r)$.

How can we estimate the spatial dependence of $D(r)$? When $r > R_0$ we expect the macroscopic value D. Passing to $r < R_0$ is equivalent to studying the migration of a reagent on a time scale $t < \tau$. Let s be the displacement of a reagent in its own tube; the quadratic mean of s is given by equation (18). The spatial displacement r that corresponds to S is of order

$$r^2 \cong |s| \delta, \tag{28}$$

where δ is the diameter of the tube section. Equation (28) expresses the fact that each tube has itself the shape of a random path. We can thus deduce the migration law (de Gennes, 1971)

$$r^4 \cong \delta^2 D_{tube} t. \tag{29}$$

We then obtain:

$$D(r) = \frac{r^2}{t} = D_{tube} \frac{\delta^2}{r^2} \quad (r < R_0). \tag{30}$$

Notice that when $r = R_0$ equation (30) recovers the correct form for macroscopic diffusion:

$$D = D_{tube} \frac{\delta^2}{R_0^2} \cong N^{-2}. \tag{31}$$

If we introduce at this point (30) into (27), we arrive at an expression which is independent of the trapping radius b:

$$k \cong 8\pi D R_0 \quad (t > \tau). \tag{32}$$

Comparing this with (24) we see that b has been replaced by R_0. The key conclusion is that, in the regime we have considered $(t > \tau)$, we can very much expect a second-order kinetics, the constant k being proportional to $N^{-3/2}$.

There exists a large amount of experimental data on these kinds of reaction, but quite some time will be required to understand them in detail: (a) the reactions are often more complex than those considered above; (b) the dependence on molecular weight has not yet been studied very systematically; (c) even within the idealized schemes that we have considered here, one expects an extraordinary variety of regimes, depending on the time scales. However, we may hope that, in the next few years, a coherent scheme will emerge for the chemical kinetics in macromolecular melts.

1.5   Additional research directions

In addition to the mechanical (rheology and adhesion) or chemical (kinetics) studies that we have briefly examined, the reptation concept finds applications in quite different sectors of polymer physics.

1. What happens to reptation when the chain under consideration is immersed in chemically identical chains with different length? This aspect is being systematically studied experimentally, yielding rather subtle theoretical situations.

2. The *crystallization* mechanism of polymers in liquid phase is still controversial, as we mentioned in section 1.1 (*Faraday Transactions*, 1979). Nevertheless, for some fast crystallization kinetics, it is possible for a chain to be extended in the amorphous phase by a 'sucking' process which is directly related to reptation (Klein, 1979).

3. What happens in a liquid of flexible chains when the chains are *branched*? From the theoretical point of view, it appears that reptation is strongly inhibited.

The above examples provide just a partial – and biased! – illustration of chain dynamics. We hope, however, that it can be useful to show the very broad spectrum of interests which converge in polymers science: from the topology of knots to how glues work . . .

2

Minimum number of aminoacids required to build up a specific receptor with a folded polypeptide chain

2.1 Introduction

From the point of view of polymer physics, globular proteins are reminiscent of certain polysoaps, where the hydrophobic part of the chain clusters in a central core, while the hydrophilic residues tend to lie on the outer surface (see Stryer, 1968). This conformation provides both *stability* and *solubility*. The crucial difference with polysoaps lies of course in the presence of *specific receptors* on the protein. A schematic representation for such a receptor is shown in Fig. 2.1. The active site directly involves a number p of aminoacids. These are linked together by comparatively long loops of the peptidic chain. As has been emphasized by Monod (1969), it is of some interest to estimate the *minimum size* required for each of these loops, when the conformation of the active site itself is prescribed. This should lead in particular to one lower bound for the molecular mass of a globular protein carrying one active site.

One possible approach to this problem would be to use Monte Carlo methods on a computer. This, however, (a) is expensive and (b) gives very little insight. Here we shall restrict ourselves to a much more modest, but explicit, calculation, based on rough statistical arguments. We deliberately neglect all the effects related to the hydrophilic/hydrophobic affinities of the aminoacids, although these effects are certainly very important. In the present chapter, we consider first a single loop,

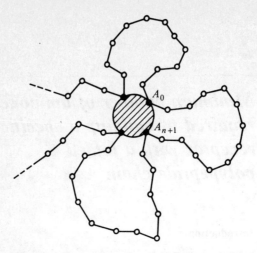

Fig. 2.1. A receptor site (hatched area) involving four aminoacids (black dots) directly. The connecting chain (white dots) comprises three loops and two open ends.

inserted in a dense proteic medium. We show that there does exist a well-defined (and rather large) minimum size for such a loop. Then we extend the argument to a set of $p-1$ loops converging towards the same active site, and show that the minimum size per loop remains essentially unaltered in this case.

2.2   The one-loop problem

2.2.1   *Statement of the problem and number of allowed conformations*

Consider a sequence $A_0, A_1, \ldots, A_n, A_{n+1}$ of aminoacids, following each other in the primary structure of the protein. We assume that A_0 and A_{n+1} belong to the active site, as shown in Fig. 2.1. Their relative positions and orientations are fixed. The other aminoacids ($A_1 \ldots A_n$) are assumed not to play a direct rôle in the receptor activity. We postulate that for a given sequence $(A_1 \ldots A_n)$ there is one (and only one) conformation of lowest energy which can be achieved. Depending on the choice of $(A_1 \ldots A_n)$ we might then have 20^n conformations for our loop. However, all of them are not realizable. First, we must retain only those conformations for which the loop under study does *not* intersect other portions of the overall protein chain: we call these the allowed conformations. Counting the allowed conformations corresponds to an excluded volume problem in a dense homogeneous medium, familiar from the theory of concentrated polymer systems (for a discussion of this concept applied to proteins, see Flory, 1961). The result (originally derived by Flory with the help of a lattice model) is that, for each unit A added to the chain, the probability of not intersecting any other part of the chain is a constant, and roughly of order e^{-1} (where e is the base of Neperian logarithms). With this estimate, the number of allowed conformations is reduced to $(20/e)^n$. Our problem is to find how many of these are compatible with the geometrical requirements imposed by the active site.

2.2.2 *Position and orientation of the end group*

To each of the $(20/e)^n$ conformations would correspond a well-defined position and orientation of the end group A_{n+1} with respect to A_0. The positions may, for instance, be labeled by the coordinates (x, y, z) of the α carbon in A_{n+1} with respect to the c carbon in A_0. Similarly the angular properties may be specified in terms of a rotation vector $(\Omega_x, \Omega_y, \Omega_z)$ relating the actual orientation of, say, the amide group in A_{n+1}, to a chosen reference state. The set $(x, y, z, \Omega_x, \Omega_y, \Omega_z)$ defines one point M in a six-dimensional space. There are $(20/e)^n$ such points M to consider. How are they distributed?

As soon as $n \gtrsim 4$ the angular variables are spread more or less at random (each of them over an interval 2π). The spatial variables, on the other hand, have a nearly gaussian distribution, even in the presence of strong excluded volume effects, provided that the medium is densely filled and homogeneous. Homopeptide chains are known to become gaussian in solution only at very large $n(\sim 30)$. But here we are interested in heteropeptide chains with a statistical distribution of sequence: such systems tend to be gaussian even for rather small $n(\gtrsim 10)$. This leads us to a distribution function $\rho(M)$ for the points M of the form:

$$\rho(M) = \left(\frac{20}{e}\right)^n (2\pi)^{-3} \left(\frac{2\pi}{3} nb^2\right)^{-3/2} \exp\left(-\frac{3x^2 + y^2 + z^2}{nb^2}\right)$$

(1)

where nb^2 is the mean square elongation of a chain containing n arbitrary aminoacids. The general aspect of this distribution is shown in Fig. 2.2.

In all that follows, we shall be interested only in *loops*, i.e. in conformations for which the end point is rather close to the starting point, or $|x|, |y|, |z| < n^{1/2}b$. In this limit $\rho(M)$ takes the simple form:

$$\rho(M) \to \rho_0 = C\left(\frac{20}{e}\right)^n n^{-3/2}b^3$$

(2)

$$C = 3^{3/2}(2\pi)^{-9/2}$$

2.2.3   *Number of 'successful' conformations*

We call successful the conformation for which A_{n+1} is indeed located as desired to build up the active site. This implies that x, for instance, be in a certain interval:

$$x_0 \leqslant x \leqslant x_0 + x$$

and that the rotations Ω_x also belong to a certain small interval:

$$\Omega_{2x} \leqslant \Omega_x \leqslant \Omega_{2x} + \Delta\Omega_x$$

(with similar inequalities for the other components). The amplitudes Δx, $\Delta\Omega_x$, etc. are imposed by the thermal fluctu-

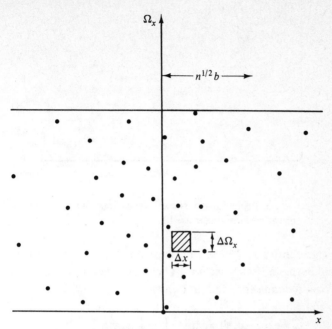

Fig. 2.2. Distribution of the positions (x) and rotation angles (Ω_x) of the terminal aminoacid A_{n+1} with respect to the first A_0, for various primary sequences ($A_1 \ldots A_n$) of the inter-connecting loop. The region which satisfies the requirements on A_{n+1} to build up the active site is hatched.

ations of the overall structure, and depend essentially on the rigidity of the protein. In what follows we shall take typically:

$$\Delta x = \Delta y = \Delta z = 0.2\,\text{Å}$$
$$\Delta\Omega_x = \Delta\Omega_y = \Delta\Omega_z = 1/10\,\text{rad} \cong 6°$$

The points M associated with successful conformations of the loop are thus all contained in a small six-dimensional volume w:

$$w = (0.2)R^3 \times 10^{-3}.$$

The probability of failure p_f is the probability that *no* point M falls into the volume w. Let us assume that, in the region of interest, the points M are distributed at random, with an

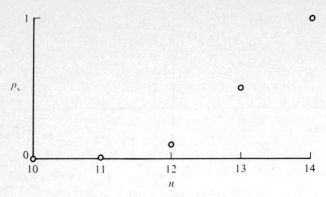

Fig. 2.3. Probability of success versus loop length. Note the abrupt variation near $n = 13$.

average density ρ_c. (We shall discuss this assumption in more detail in the following section.) Then p_f is given by a standard Poisson formula for random events:

$$p_f = \exp(-\rho_c w) = \exp\left[-\left(\frac{20}{e}\right)^n \frac{Cw}{n^{3/2}b^3}\right]$$

The probability of success is $p_s = 1 - p_f$. The dependence of p_s on n, as deduced from this equation, is shown in Fig. 2.3. Because of the 'doubly exponential' character of equation (3), p_s is essentially 0 for all values of n below a certain threshold n_c, and $p_s \cong 1$ for $n > n_c$. Taking $b = 4$ Å, and the aforementioned value for w, we are led to $n_c \cong 13$. It turns out in fact that n_c is remarkably insensitive to the exact values chosen for b and w (the dependence being only logarithmic).

2.3   Discussion

2.3.1   *Could the same conformation be obtained with more than one sequence?*

One basic assumption underlying equation (3) is that, in the region around w, the points M are spread out at random, without any strong correlation between them. This might be incorrect in some cases.

Consider, for instance, the 'stereochemical code' proposed for aminoacids residues by Liquori (1968). In this model, the aminoacids may have only five different conformations (i.e. different sets of angles ϕ and ψ in the conventional notation: see Edsall *et al.* (1966). If this rule was strictly obeyed, the number of distinct conformations would be reduced from 20^n to 5^n; the same point M_0 would correspond to a large number of primary sequences. Our calcuation would then lead to a much larger value (~ 30) for n_c.

We believe, however, that our earlier estimate for n_c is the correct one, for the following reason: the angles ϕ and ψ for each aminoacid do have some small deviations from the idealized values selected by Liquori. Thus, the many sequences, which, in a strict version of the code, would converge to the same point M_0, will in fact lead to a *cluster* of points M centered around M_0 (Fig. 2.4). Are these clusters well separated from each other, or do they overlap? The calculation of the previous section applies only when they overlap, since then the overall density of points M becomes uniform.

To answer this question, let us first estimate the size of one cluster. A spread ε in the angles for each aminoacid implies for the orientation of the final group A_{n+1} an uncertainty:

$$\delta\Omega_x \sim \delta\Omega_y \sim \delta\Omega_z \sim \varepsilon n^{1/2}.$$

As regards the position of A_{n+1}, since the overall loop size is of order $n^{1/2}b$, a change ε in the angles for *one* aminoacid will displace x, y, and z by $\sim \varepsilon n^{1/2}b$. The cumulative effects of n aminoacids will give a result $n^{1/2}$ larger, i.e.:

$$\delta x \sim \delta y \sim \delta z \sim n\varepsilon b.$$

Thus the volume of one cluster is approximately:

$$\Delta = \varepsilon^6 n^{9/2} b^3.$$

On the other hand, the number of clusters per unit volume in six-dimensional space, is given by the analog of equation (2):

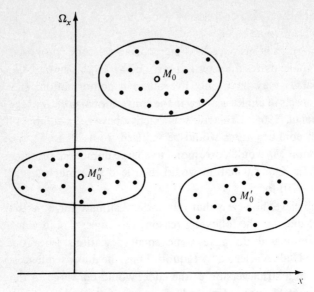

Fig. 2.4. Clustering effects in the (x, Ω_x) diagram. If the stereochemical code of Liquori was strictly obeyed, various primary sequences would give the same point M_0. Small deviations from the code spread these points in a cluster around M_0.

$$\tilde{\rho}_c = Cb^{-3}n^{-3/2}.$$

The overlap will be large if:

$$\tilde{\rho}_c \equiv C\varepsilon^{-6}n^3.$$

From the experimental distributions of points in the $(\phi\psi)$ plane for proteins of known structure (see, for instance, Ramachandran, 1968) we estimate $\varepsilon \cong 0.2$ rad (or $12°$). Then $\tilde{\rho}_c\Delta$ becomes larger than 1 as soon as $n > 15$. Thus (if our guess for ε is correct), this 'clustering' effect will at most increase n_c by two units, and is not too important.

2.3.2   Effects of the protein surface

The receptor site has to be close to the surface, and the loops must reach the surface sometimes. As a model system, we may take a planar surface, and restrict the loop paths to one half space limited by this plane (the receptor site being treated as a point on the surface). The result of such a model is simply to modify slightly the coefficient C in equation (2), and the effect on n_c is very weak.

Of course, the surface will also influence the hydrophilic/hydrophobic balance of the loop sequence, and this should be included in a more refined calculation.

2.4   The many-loop problem

When the receptor site under consideration involves directly a number $p > 2$ of aminoacids (as it probably does in most cases), we have $p - 1$ loops. Does the same value of n_c apply to all of them?

At first sight the answer would seem to be no, since the first loop restricts the amount of phase space available for the second one, etc. But we believe that in fact the answer is yes, for the following reasons.

(a) The Flory factor $\exp(-n)$ does contain the effect on one loop, of all the material present, and in particular of the other loops.

(b) It is of course true that, if p is comparatively large, many loops will have to converge towards the same small region defined by the active site: this implies a reduction in entropy since the loop will have to stretch radially as shown in Fig. 2.5. A reduction in entropy means an unfavorable factor. However, we shall now estimate this reduction and show that it is weak.

Assume for instance that the $p - 1$ loops fill completely a spherical region of radius R around the active site. Apart from numerical coefficients, the volume of this region will be

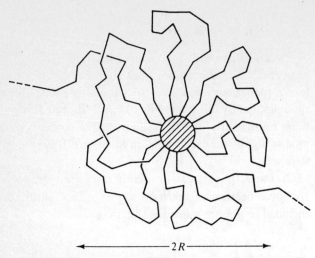

Fig. 2.5. When the number of loops is large, their paths near the active site are stretched, with a corresponding reduction in entropy.

$$R^3 \sim (n+1)(p-1)v \sim npv$$

where v is the average volume per aminoacid, and n the average chain length. Each loop is stretched over a distance of order R and suffers a reduction in entropy R^2/nb^2 (in dimensionless units). This corresponds to a reduction factor in the number of allowed conformations per loop, of order:

$$\exp(-R^2/nb^2) \sim \exp(-p^{2/3}n^{-1/3})$$

(assuming $v \sim b^3$). For the values of p and n of interest here, this factor is not much smaller than unity, and it may be omitted completely.

Thus, at least to a first approximation, our estimate of n_c should apply to the many-loop problem. Since there are $n-1$ loops, each of them involving at least n_c+1 aminoacids, the total number of aminoacids required to build up the receptor site may not be smaller than $(p-1)(n_c+1)$. Of course, this is only a lower limit: in particular, the requirements of stability, rigidity, and solubility will often impose longer chains.

3

Dry spreading of liquids on solids

3.1 General aims

The wetting properties of liquids against a solid surface are important for many practical purposes. A vast amount of knowledge has been collected by Zisman *et al.* (1964), mainly for the case of partial wetting, where contact angles can be measured, and related to interfacial energies. Situations of complete wetting are equally important, but have suffered from a certain lack of experimental observables. In the present paper, we shall concentrate on the special case of complete, 'dry', spreading, where the vapor pressure of the liquid is negligible, and the solid–gas interface does not carry any adsorbed layer. The essential physical parameter controling this process is the spreading coefficient, S, introduced long ago by Cooper and Nuttal (1915).

$$S = \gamma_{SO} - \gamma_{SL} - \gamma \tag{1}$$

where γ_{SO} and γ_{SL} are the interfacial energies between solid/air and solid/liquid, while γ is the surface tension of the liquid. We are interested here in cases of positive S, and we shall try to understand first the final state of a spreading droplet (Section 3.1). It was commonly stated that the final state is a monomolecular film: but we shall see that this need not be true at small S, where the final state should be a relatively thick 'pancake'. In Section 3.3 we review some major features of the dynamics of the dry spreading. Finally, certain special cases (polymers, superfluids) are presented in Section 3.4.

3.2   The final 'pancake'

An essential observation, going back to the early work of the
Russian school (Derjagin, 1955) is that thin liquid films are very
sensitive to long-range forces. This can be described in terms of
a long-range energy (per cm^2) $P(e)$, such that $P(e \to \infty) \to 0$. The
energy $P(e)$ depends both on the liquid and on the solid, and on
the film thickness e. Equivalently, one may describe the long
range forces through a 'disjoining pressure':

$$\Pi(e) = -\frac{\mathrm{d}P}{\mathrm{d}e} \tag{2}$$

For the simplest example (a pure Van der Waals fluid)

$$P(e) = \frac{A}{6\pi e^3}, \tag{3}$$

A being a Hamaker constant. For more complex fluids, such as
water, the exact structure of $P(e)$ is not fully understood at
present, but there exist some reliable data for certain solid
substrates (Pashley, 1980).

Consider now a macroscopic droplet which has spread out
and covered an area $\mathscr{A}$ has two antagonist terms

$$f = -S\mathscr{A} + \mathscr{A}P(e). \tag{4}$$

We must minimize f at constant volume $\Omega = e\mathscr{A}$. The result is
an implicit equation for e (Joanny and de Gennes, 1984):

$$S = e\Pi(e) + P(e) \tag{5}$$

The resulting $e(S)$ is a *decreasing* function of S. For instance, in
Van der Waals case (equation (3))

$$e = a\left(\frac{3\gamma}{2S}\right)^{1/2} \tag{6}$$

where a is a molecular size, defined by

$$a^2 = \frac{A}{6\pi\gamma}. \tag{7}$$

If $S/\gamma \sim 1$ equation (6) gives a thickness comparable to a, in agreement with the common creed. But if $S/\gamma \ll 1$ we do expect a relatively thick 'pancake' ($e \sim 100\,\text{Å}$). This has many implications.

(a) Equations (5) or (6) give us an interpretation of the Cooper–Nuttal rule. If, for instance, we want to spread an insecticide on leaves (the problem originally considered by Coopes and Nuttall), choosing a large S will give a thin film, and thus a much better coverage.

(b) If the final thickness e is measured (by ellipsometry or by some equivalent technique), equations (5) and (6) give us information on the spreading coefficient. In the future, these measurements (at positive S) may become just as relevant as the measurements of contact angles (at negative S).

Let us end this section by a technical comment, concerning the edge of the pancake. The shape of the edge can be calculated simply by including capillary energies, Van der Waals forces, and the S parameter (Joanny and de Gennes, 1984; de Gennes, 1985a). The results are shown in Fig. 3.1. The main parameter of interest is the healing length ξ_h which describes the horizontal size of the edge region, and which turns out to be

$$\xi_h \sim e^2/a \tag{8}$$

for Van der Waals forces. The edge has a line energy $\mathcal{T}$ which is of order (omitting logarithmic factors)

$$\mathcal{T} \sim S\xi \sim \gamma a \tag{9}$$

Fig. 3.1. The final 'pancake' profile in dry spreading. The thickness $e(S)$ is constant except in an edge region of width ξ_h.

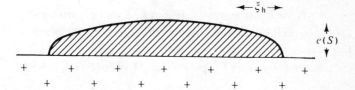

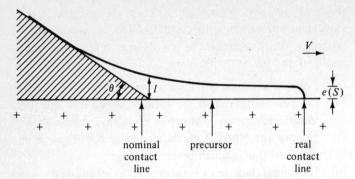

Fig. 3.2. Detailed form of the advancing profile in dry spreading of a completely wetting fluid. The cross-over thickness l is of order a/θ.

These considerations are important for a discussion of dry spreading on a random surface (i.e., when S is slightly modulated from point to point): for more details see de Gennes (1985b) and Joanny (1985).

3.3   The dynamics of dry spreading

The basic situation is illustrated in Fig. 3.2. The advancing fluid displays a certain dynamic contact angle θ, which depends on the contact line velocity U. Three separate experiments by Hoffmann, Marmur and Tanner indicate that for various liquids, and various solids, the relation between U and θ is universal

$$U = (\text{constant})\frac{\gamma}{\eta}\theta^3 \quad (e \ll 1) \tag{10}$$

where η is the liquid viscosity. It is at first very surprising that the (U, θ) relation does not involve the spreading coefficient S. For instance, if we think in terms of irreversible thermodynamics, the dissipation $T\overset{\circ}{\Sigma}$ (per unit length of line) is the product of a flux (U) by a force F, which is the 'non-compensated Young force'

$$F = \gamma_{SO} - \gamma_{SL} - \gamma\cos\theta \tag{11}$$

$$\cong S + \tfrac{1}{2}\gamma\theta^2 \quad (\theta \ll 1) \tag{12}$$

In typical conditions ($\theta = 10^{-2}$, $S/\gamma = 10^{-1}$) the angular correction $\tfrac{1}{2}\gamma\theta^2$ amounts only to a very small fraction (10^{-3}) of the main force S. It is thus very surprising that S does not show up in equation (10).

The explanation lies in the existence of a *precursor film* ahead of the macroscopic droplet: this film was detected long ago by Hardy and its properties are analyzed in a recent review by Marmur (1983). On the theoretical side it is clear that these films depend on long-range forces: a detailed analysis has been carried out by two groups: (a) Teletzke, Davis and Scriven mostly considered wet spreading and made a large use of computer calculations; (b) Hervet and the present author (de Gennes, 1985a; Hervet and de Gennes, 1984) concentrated on the dry spreading of simple Van der Waals films (see Fig. 3.2).

The main conclusion is that the large free energy S described in equation (12) is entirely burned in the precursor film while the remaining energy ($1/2\gamma\theta^2$) is spent into macroscopic viscous flow. The flows in an advancing wedge were discussed long ago by Huh and Scriven (1971): for small θ, the corresponding dissipation can be derived very simply using a lubrication approximation, and is

$$T\mathring{\Sigma}_{\text{macro}} = kU^2/\theta \tag{13}$$

where k is nearly constant (a logarithmic function of θ). Equating $T\mathring{\Sigma}_{\text{macro}}$ to the work done by $1/2\gamma\theta^2$ we arrive immediately at the experimental law (10).

This has a number of practical consequences:

(a) we see that the dynamics is finally independent of S (for $S > 0$). Thus the Cooper–Nuttal law cannot be explained by dynamic effects, and depends entirely on properties of the final equilibrium state (equation (5));

(b)   there are many pitfalls: for instance it is completely incorrect to equate the larger dissipation term SU to the macroscopic losses $T\overset{\circ}{\Sigma}_{macro}$. Mistakes of this type have occurred in the literature.

The above discussion was restricted to horizontal spreading, with negligible gravity effects. The case of vertical spreading is also of some interest (de Gennes, 1984). For instance, if a vertical plate is pushed down at *low* velocities U inside a vessel of wetting liquid, the fluid still builds up a stationary wetting fluid along the plate. In the lower part of this film, there is a quasi-static equilibrium between disjoining pressure and hydrostatic pressure, giving a thickness $e \sim h^{-1/3}$ (where h is the altitude) for Van der Waals fluids. But in the upper part of the film, the thickness $e(h)$ decreases more rapidly, very much as in the horizontal case $(e \sim h^{-1})$. These considerations may be of some interest in soil physics.

3.4   Special systems

(1) *Polymer melts* have very special spreading laws, as pointed out in particular by Ghiradella, Radigan and Fritsch (1978). (See also Sawicki, 1978; Arslamov *et al.*, 1971.) Certain regimes (not all!) might possibly be understood from a remark made in 1979 (de Gennes, 1979): namely that polymers flowing near a smooth surface tend to have a very anomalous slippage. This idea is supported by some early experiments of B. Maxwell using a transparent extruder (Galt and Maxwell, 1964) and also by more recent rheometric studies of Burton *et al.* (1983). The consequences of this strong slippage have been worked out (Brochard and de Gennes, 1984): the predicted transient profiles for a droplet should show a characteristic 'foot'.

(2) *Superfluid helium 4* is an interesting candidate for spreading experiments, but very little is known here on the theoretical

side. Joanny (1985) considered the very special case $S = 0$, and constructed one exact self-similar solution for macroscopic spreading of a droplet. In this case, at large times t, all the available capillary energy is transformed into kinetic energy, and the contact line advances at a constant velocity U. But, for most solids, we expect to have $S > 0$ and not $S = 0$. This case of positive S is very mysterious: we do not know where the energy S can go! Three choices are possible:

(a) S could be transformed into a kinetic energy for macroscopic flow. This would then lead to a very fast growth ($U(t)$ increasing exponentially with time). But this assumption is open to some doubt: for viscous liquids, as pointed out in section 3.3, it is completely wrong to think that S can drive the macroscopic motion. A similar feature might hold for superfluid helium;

(b) S could be used in the form of capillary waves;

(c) S could be used to create vortex lines in the precursor film of helium.

We clearly need an extended program of experiments to tell us what really happens with helium spreading on a solid!

4

An elastic theory of drag reduction

4.1   Introduction

Flexible polymers in dilute solution can reduce turbulent losses very significantly (Lumley, 1969). The main (tentative) interpretation of this effect is due to Lumley (1973). He emphasized that remarkable viscoelastic effects can occur only when certain hydrodynamic frequencies become higher than the relaxation rate of one coil $1/T_z$ (the 'time criterion'). He then proposed a crucial assumption: namely that, in regions of turbulent flow, the solution behaves as a fluid of *strongly enhanced viscosity*, presumably via regions of elongational flow. On the other hand, Lumley noticed that – for turbulent flow near a wall – the viscosity in the laminar sublayer near the wall should remain *low*: this last observation does agree with the viscometric data on dilute linear polymers in good solvents, which show shear thinning (Graessley, 1974). Starting from the above assumptions, and performing a careful matching of velocities and stresses beyond the laminar layer, Lumley was able to argue that the overall losses in pipe flow should be reduced.

This explanation has been rather generally accepted. However, it is now open to some question: in recent experiments with polymer injection at the center of a pipe, one finds drag reduction in conditions where wall effects are not involved (McComb and Rabie, 1979; Bewersdorff, 1982, 1984).

This observation prompted Tabor and the present author to try a completely different approach (Tabor and de Gennes, 1986): namely to discuss first the properties of homogeneous, isotropic, three-dimensional turbulence *without any wall effect*, in the presence of polymer additives. This 'cascade theory' is described in Section 4.2. The central idea is that polymer effects at small scales (high frequencies) are not described by a viscosity, but by an *elastic modulus*. The general notion of elastic behavior at high frequencies is classical for molten entangled chains (Graessley, 1974). It may also be important for dilute polymers. In our approach the viscosity effects are mostly trivial, and we do not even discriminate between solvent viscosity and solution viscosity.

In Section 4.3 we return to wall turbulence, and try to set up a modified version of the Lumley approach, where, at each distance y from the wall, we have a cascade, but it is truncated elastically. This gives a law for the minimum eddy size r^{**} versus distance y which is qualitatively different from Lumley's viscous effect. But the net result is still an enhancement of the intermediate 'buffer layer'. We expect drag reduction from this, although we have not carried out the detailed analog of Lumley's matching.

In Section 4.4 we list some more general systems which can show drag reduction on turbulent flow. Some of the systems are dominantly elastic while others are probably dominantly viscous.

Our whole discussion is very qualitative. But, even at this modest level, it leads to a surprisingly rich classification of possible cascades and flows. For instance, in bulk turbulence, we have three control parameters: (a) the dissipation per unit mass ε; (b) the polymer chain length, or equivalently the number of monomers per chain N; (c) the monomer concentration c (or the number of coils/cm^3, $c_p = c/N$).

This three-dimensional parameter space can be split into regions where different 'scenarios' for the cascade should occur. The identification of these scenarios is a natural aim for future experimental research.

4.2   The cascade theory

4.2.1   The time criterion

Our starting point is the classical view of Kolmogorov (see Tennekes and Lumley, 1972) for homogeneous, isotropic, three-dimensional turbulence.

At each spatial scale (r) there is a characteristic fluctuating velocity $U(r)$, related to (r) by the condition

$$\frac{U^3(r)}{r} = \varepsilon = \text{constant}. \tag{1}$$

We must compare the characteristic frequencies $U(r)/r$ to the Zimm relaxation rate of one coil (Stockmayer, 1976; de Gennes, 1979)

$$\frac{1}{T_z} \cong \frac{kT}{\eta_0 R^3}, \tag{2}$$

where k is the Boltzmann constant, T the temperature, η_0 the solvent viscosity, and R the gyration radius of the coil at rest. We focus our attention on linear, flexible, neutral polymers in good solvents, where the Flory law holds (Flory, 1971; de Gennes, 1984)

$$R \cong N^{3/5}a, \tag{3}$$

a being a monomer size, and N the number of monomers per coil.

At large scales r, the hydrodynamic frequency U/r is smaller than $1/T_z$. But, if we go down in scale, we may reach a value $r = r^*$ where the two frequencies become equal. Thus

$$U(r^*)T_z = r^*. \tag{4}$$

Solving the coupled equations (1) and (4) we arrive at

$$r^* = (\varepsilon T_z^3)^{1/2} \quad (\sim N^{2.7}\varepsilon^{1/2}), \tag{5}$$

$$U^* = (\varepsilon T_z)^{1/2} \quad (-N^{0.9}\varepsilon^{1/2}). \tag{6}$$

Note that r^* (and U^*) depend on molecular weight, but not on concentration.

Another parameter of interest for our discussion will be the Reynolds number computed *at the scale* r^*, namely

$$\mathrm{Re}^* = \frac{U^* r^*}{v} = \frac{v T_z^2}{v} (\sim N^{3.6}\varepsilon), \tag{7}$$

where $v = \eta_0/\rho$ is the kinematic viscosity, ρ being the fluid density.

The condition (4) defining r^* is the natural expression of Lumley's time criterion (Lumley, 1973). Most interesting viscoelastic effects will occur only at frequencies higher than $1/T_z$, or equivalently at scales $r < r^*$.

4.2.2 The passive range

If our solute macromolecules are very dilute, their reaction on the flow pattern is weak. Thus we expect that there exists a certain interval $r^* > r > r^{**}$ where eddies of size r are still described by the Kolmogorov cascade, but where the polymer begins to undergo strong distortions.

Information from laminar flows

Let us concentrate first on *elongational flows*. Two regimes have been probed in some detail.

(i) *Constant shear rate* $\dot{\gamma}$ (Fig. 4.1). Here one expects that the coils are essentially unperturbed when $\dot{\gamma} < 1/T_z$ and that they are strongly elongated when $\dot{\gamma} > 1/T_z$ (Hinch, 1977; de Gennes, 1974; 1984). This sharp coil-stretch transition has been observed in an important series of experiments by Keller and coworkers (Pope and Keller, 1978; Odell and Keller, 1986). We might think at first sight that this transition should show up in turbulent flows and bring in some important nonlinear effects.

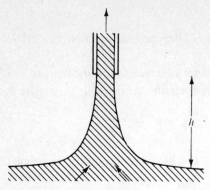

Fig. 4.1. The tubeless siphon: a dilute solution of long flexible polymers can be sucked up over large intervals $h(\sim 20\,\text{cm})$. This shows the dramatic effects of the polymer-induced stresses in longitudinal shear flows.

We shall argue, however, that this is not correct: for the situations of interest, where $\dot{\gamma}$ (as seen by the molecule) is rapidly varying in time, the coil stretch–transition disappears completely.

(ii) *Variable shear rates.* Two examples are shown in Figs. 4.2*a,b*: in Fig. 4.2*a* we have a duct with a periodic modulation of the cross section. In Fig. 4.2*b* we consider the converging flow towards a very thin ($\sim 500\,\text{Å}$) capillary. The main conclusion, obtained first from detailed calculations by Daoudi and Brochard (1978) for a coil under periodic modulations, is the following: whenever the modulation frequency is higher than the Zimm relaxation rate, the coil *follows passively the deformations of the local volume element.* The dimensionless elongation λ of the coil is entirely fixed by the flow.

Of course there are still some local modulations in the coil shape: more specifically, if we call ω the modulation frequency, we can define subunits of p monomers such that

$$\frac{1}{T_z(p)} = \omega,$$

$$T_z(p) = \frac{\eta_0 a^3 p^{1.8}}{kT}, \tag{8}$$

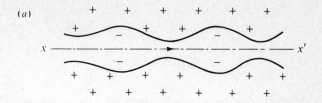

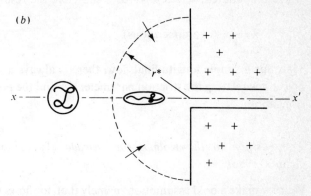

Fig. 4.2. Two (approximate) examples of longitudinal shear flows. (a) Tube with periodic constriction; (b) entry of a capillary. In both cases the molecules which lie exactly on the axis of symmetry (xx') experience a purely longitudinal shear.

where $T_z(p)$ is the Zimm type of the subunit. Inside each subunit we still have some relaxation, but at larger scales the coil deforms affinely.

These considerations have been transposed long ago (Daoudi and Brochard, 1978) to the converging flow of Fig. 4.2b. Here the shear rate at a distance r from the entrance point is of order

$$\dot\gamma(r) = \text{const.}\ r^{-3}. \tag{9}$$

Viscoelastic effects occur for $r < r^*$, where:

$$\dot\gamma(r^*) = 1/T_z. \tag{10}$$

At distances $r > r^*$ the coils are not deformed. At distances $r < r^*$ they deform affinely, and their elongation is

$$\lambda = \left(\frac{r^{*2}}{r}\right)^2 \text{ (three dimensions).} \tag{11a}$$

A similar discussion can be given for a two-dimensional flow, where the fluid converges towards a slit. Here the result is

$$\lambda = \frac{r^*}{r} \text{ (two dimensions).} \tag{11b}$$

Thus, for a simple longitudinal flow, there is always a simple power law relating the striction parameter r^*/r and the polymer elongation.

Transposition to the Kolmogorov cascade (Tabor and de Gennes, 1986)

We now make a bold assumption: namely that, for flows which are admixtures of longitudinal shear and simple shear, and which are turbulent, there remains a power law between polymer elongation and spatial scale

$$\lambda(r) = \left(\frac{r^*}{r}\right)^n, \tag{12}$$

where n is an unknown exponent. The extreme case quoted in equation (11) suggests that $n < 2$. In practical discussions we shall attempt to use $n = 1$ and $n = 2$ as possible values.

It may be worth while to return here to the definition of molecular elongation. For one particular coil, the dimensionless elongation $\lambda_{(1)}$ can be constructed from the radius of gyration $\tilde{R}$ in the distorted state

$$\lambda_{(1)}^2 = \frac{\tilde{R}^2}{R^2}, \tag{13}$$

where R is the radius at rest (equation (3)). For an ensemble of coils in a turbulent flow, we should select the coils which belong, in real space, to eddies of size r (and which do not belong to any smaller eddy). Then the average of $\lambda_{(1)}$ over this population is what we call λ. One immediate question concerns the distribution of $\lambda_{(1)}$ values within the population. In the present, naive, approach, we assume that this distribution is reasonably *narrow*, so that, for instance, the average of the squares is given by

$$\langle \lambda_{(1)}^2 \rangle = k_2 \lambda^2 \tag{14}$$

with a constant k_2 which is independent of r, and of order unity. It may well be that this assumption of narrow distributions is not satisfactory, and that independent exponents $n_1 n_2 \ldots$ would be required to describe the successive moments of the elongation. But, at our present level of ignorance, we shall ignore this complication.

To summarize: we expect that any coil, located in eddies of size $r < r^*$, will follow passively the surrounding volume element, and will deform accordingly. We postulate a scaling law describing this effect (equation (12)) in terms of a single exponent n.

This simple behavior, with affine deformation, and without significant reactions of the coils on the flow, will hold in a finite interval of spatial scales $r^* > r > r^{**}$. We call this interval the *passive range*.

4.2.3   The first scenario: semi-stretched chains

Stresses in a partly stretched state

Let us now consider the reaction of the polymers on the flow: when our coils are stretched by a factor λ, a certain elastic energy is stored in each of them. We shall, for the moment, assume that the coils are significantly stretched ($\lambda \gg 1$) but that they are still far from full extension – the deformed size $\tilde{R}$ is still much smaller than the contour length Na - or returning to equations (3)–(13):

$$1 \ll \lambda \ll N^{2/5}. \qquad (15)$$

Since $N \sim 10^4$–10^5 in typical experiments, we may go up to $\lambda \sim 100$.

What is the elastic energy of a coil in this regime? In the harmonic approximation, it would be proportional to $(\lambda - 1)^2$ (or equivalently to λ^2, since $\lambda \gg 1$). However, for coils in good solvents, the harmonic approximation is not very good: the shape, and the monomer repulsions, change with λ. This has been analyzed by Pincus (1976).

The final result is an anharmonic energy

$$F_1 \cong kT\lambda^{5/2} \quad (1 \ll \lambda \ll N^{2/5}) \qquad (16a)$$

or a free energy per unit volume

$$F_{e1} = \frac{c}{N} kT\lambda^{5/2} = G\lambda^{5/2}. \qquad (16b)$$

where G has the dimensions of one elastic modulus, and is linear in concentration. Equivalently the restoring force on one spring is

$$f_1 \cong \frac{kT}{R} \lambda^{3/2} \qquad (17a)$$

and the stress due to c/N springs/cm^3 is

$$\tau \cong \frac{c}{N} f_1 \lambda R \cong F_{e1}. \qquad (17b)$$

The elastic limit r_1^{**}

Whenever τ is much smaller than the Reynolds stresses ρU^2, the reaction of the polymer on the flow is negligible. If we go towards smaller and smaller scales, λ and τ increase, while the local Reynolds stress $\rho U^2(r)$ decreases. Thus, at a certain scale

r_1^{**}, the two stresses become equal.

$$G[\lambda(r_1^{**})]^{5/2} = \rho U^2(r_1^{**}). \qquad (18)$$

Using the Kolmogorov formula (2.1) this leads to

$$\frac{r_1^{**}}{r^*} = X^v, \quad v = \left(\frac{5n}{2} + \frac{2}{3}\right)^{-1}, \qquad (19)$$

where we have introduced a dimensionless parameter

$$X \equiv \frac{G}{\rho U^{*2}} (\sim cN^{-2.8}\varepsilon^{-1}). \qquad (20)$$

X is a natural measure of concentration effects. Consider a typical case with $U^* = 1\,\text{m/s}$, $\rho = 1\,\text{g/cm}^3$, $ca^3 = 10^{-4}$, $a = 2\,\text{Å}$, $T = 300\,\text{K}$, $N = 10^4$. Then $G = 60\,\text{erg/cm}^3$ and $X = 6 \times 10^{-3}$. Depending on our choice of n, the exponent v might be in the range $\frac{1}{3} - \frac{1}{6}$.

Comparison with the Kolmogorov limit

Can we effectively go down to the elastic limit r_1^{**}, retaining the inertial cascade all the time? The Kolmogorov scheme is always truncated by viscous dissipation at a scale r_z defined by (Tennekes and Lumley, 1972):

$$\frac{r^*}{r_z} = (\text{Re}^*)^{3/4}. \qquad (21)$$

The elastic limit is observable only if $r_1^{**} > r_z$. Comparing equations (19) and (21) we find that this condition is equivalent to

$$X > (\text{Re}^*)^{-3/4v}. \qquad (22)$$

The condition (22) has no counterpart in the Lumley scheme, where drag reduction was expected to occur at arbitrarily low polymer concentrations. Here, we do find a minimum X, or, equivalently, a concentration threshold c_m, below which the polymer should have no visible effect. Using equations (20) for

X and (7) for Re^*, we can find how the threshold concentration c_1 depends on N and ε:

$$c_m \sim N^{2.8 - 2.7/v} \varepsilon^{1 - 3/4v}. \tag{23}$$

Since $1/v$ is expected to be in the range 3–6, c_1 should be a strongly decreasing function of N. It may well be that, for long chains, c_m is extremely small and practically invisible. But systematic experiments at variable N might detect the threshold c_m.

Ultimate fate of the turbulent energy

At the scales $r = r^{**}$, the liquid should behave like a strongly distorted rubber, carrying elastic waves (longitudinal and transverse) with comparable kinetic and elastic energies.

At the scales $r < r^{**}$ inertia nonlinearities are not, by themselves, able to generate smaller structures. But the *elasticity* is also nonlinear, and may have the ingredients required to produce shock waves in the 'rubber'. This process may imply a further thinning of scales, down to the natural width of a shock front. This scheme is completely conjectural: we may deal with a sea of rarefaction waves and shock waves, with very peculiar couplings between them. Thus, we do not know the ultimate fate of the turbulent energy.

On the whole, it is tempting to assume that the formation of new eddies is strongly restricted for $r < r^{**}$. Using the description of equations (18)–(23), this would then lead to a truncation in the cascade at $r = r_1^{**}$ (the index 1 stands for the 'first scenario' with partly elongated chains, which was the only one discussed above).

The result is

$$r_1^{**} = r^* X^v, \tag{24}$$

$$r_1^{**} \sim N^{2.7 - 2.8v} \varepsilon^{1/2 - v} c^v. \tag{25}$$

We expect at last a qualitative change, and possibly a truncation, of the cascade at $r = r_1^{**}$. Note that r_1^{**} should increase rapidly with N.

Recent experiments on pipe turbulence or with planar mixing layers (Ambari and Guyon, 1982; Scharf, 1985), using strophometry or laser Doppler anemometry, do suggest that polymer additives can suppress certain small scales. But detailed proposals, such as equation (25), for the truncation, remain to be checked.

4.2.4   The second scenario: strongly stretched chains

Let us return to the passive regime, and assume that we can reach very small scales r, so that the polymer under flow may become fully elongated. Returning to equation (15), we see that this corresponds to $\lambda = \lambda_{max} \cong N^{2/5}$. Inserting this value into the scaling law (12), we arrive at a certain characteristic scale

$$R_2^{**} = R^* N^{-2/(5n)}. \tag{26}$$

We call r_2^{**} the stretching limit. The meaning of r_2^{**} is made more apparent by the construction of Fig. 4.3 which is a log–log plot of the stresses as a function of the striction ratio r^*/r. When λ gets very close to λ_{max}, the stresses tend to diverge (detailed laws for this have been constructed on simple molecular models): this fixes r_2^{**}.

In this second scenario the behavior at scales r smaller than r^{**} is probably rather different: viscous forces are very important when we have rod particules immersed in a fluid; it may well be that viscosity takes over at $r = r_2^{**}$.

4.3   Flow near a wall

4.3.1   A reminder about pure fluids (Tennekes and Lumley, 1972)

Wall turbulence is characterized by a velocity U_τ such that the wall stress is equal to ρU_τ^2. The average velocity profile for a

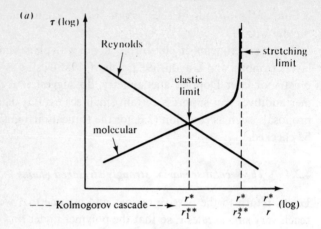

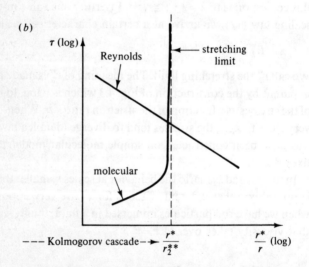

Fig. 4.3. Reynolds stresses and molecular stresses in a Kolmogorov cascade with a dilute polymer solution. (*a*) In the first scenario, the Kolmogorov cascade is truncated at a certain scale r_1^{**}: at this scale, the chains are only partly stretched; (*b*) in the second scenario, the truncation occurs (at a scale r_2^{**}) where the chains are totally stretched.

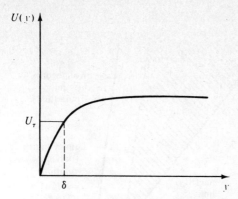

Fig. 4.4. Average velocity profile in the wall turbulence of a pure fluid (qualitative). y is the distance from the wall, and $\delta = v/U_\tau$ is the thickness of the laminar sublayer. At $y \gg \delta$ the profile is logarithmic.

pure fluid is represented in Fig. 4.4: it is linear very close to the wall, in a laminar sublayer of size

$$\delta = \frac{v}{U_\tau}; \tag{27}$$

then, at long distances from the wall, it is logarithmic. At any distance $y(\gg \delta)$ from the wall we find eddies with wave vectors k lying between two limits: the largest size (or the smallest wave vector k_{min}^{-1}) is given by the distance y itself; the smallest size (or the largest wave vector k_{max}^{-1}) is given by a Kolmogorov limit r_z similar to equation (21). The main difference is that now the dissipation ε is a function of the distance to the wall.

$$\varepsilon \cong U_\tau^3/y. \tag{28}$$

A formula, equivalent to (21), for $r_z(y)$ is

$$r_z = \left(\frac{v^3}{\varepsilon}\right)^{1/4} = \delta^{3/4}y^{1/4}. \tag{29}$$

Thus the eddies expected around the observation point (y) have wave vectors in the range

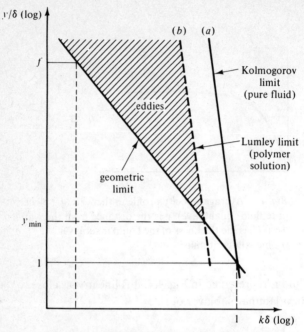

Fig. 4.5. The distribution of eddy wave vectors k at various distances y from the wall. (*a*) Pure fluid: the smallest eddies are defined by the viscous limit of Kolmogorov. (*b*) Polymer solution in the Lumley scheme: the limit is shifted to the left but the slope of the limiting line remains the same (-4).

$$\frac{1}{y} < k < \frac{1}{\delta^{3/4}y^{1/4}}. \tag{30}$$

We call the lower limit the geometrical limit, and the upper limit the viscous limit. The range of turbulent eddies is shown in Fig. 4.5.

4.3.2   *The Lumley model for polymer solutions* (Lumley, 1973)

Lumley kept the viscosity in the laminar sublayer at its Newtonian value, but he assumed an increased viscosity η_1 in the turbulent regions. This implied that, at any distance y from the wall, the Kolmogorov limit $r_z(y)$ was shifted upwards; in the log–log plot of Fig. 4.5, the new viscous limit is represented by a broken line, parallel to the original limit (slope -4 according to equation (29)). The higher the concentration, the higher the shift of the viscous limit.

The net result is a shrinkage of the turbulent domain: beyond the laminar layer, we now find a buffer layer, of size increasing with the polymer concentration. It is then natural to expect that the turbulent losses will be *reduced*; Lumley (1973) gave a detailed argument to show this. Note that the whole effect occurs at arbitrarily low c: as soon as we add some polymer, the viscous limit shifts.

4.3.3   A modified version in the first scenario

Let us assume that the discussion of Section 4.2.3 holds: the chains are partly stretched, but never fully extended, and there is an elastic limit r_1^{**}, given, in terms of the local energy dissipation ε, by equations (21), (24), (25). Inserting equation (28) for $\varepsilon(y)$, this gives

$$r_1^{**}(y) \sim U_r^{3(1/2-v)} y^{v-1/2-2.8v} c^v \tag{31}$$

Let us further assume that no singular dissipation occurs at scales smaller than the elastic limit. Then we are led to a modified Lumley construction, shown in Fig. 4.6. There is an elastic limit, described by the broken line, and the slope of this line is reversed in sign. A very low c, the new limiting line intersects the geometrical limit at thickness y smaller than the laminar sublayer δ: in this regime, we expect no macroscopic effect. But, beyond a certain threshold c_o (o stands for: onset), the turbulent domain is actually truncated. The scaling structure of c_o can be extracted from equation (5), writing $r_1^{**}(\delta) = \delta$. c_o is the wall analog of the concentration c_m introduced in our discussion of homogeneous turbulence. In many practical cases,

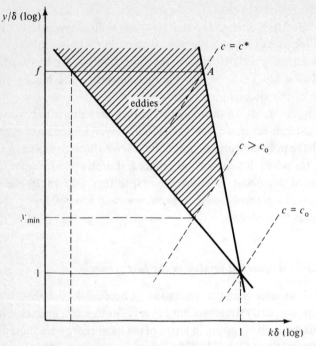

Fig. 4.6. The plot of Fig. 4.5, transposed to the present model, in the first scenario. The maximum wave vector of turbulent eddies can depend upon an elastic limit: note the difference in slope with Fig. 4.5. At very low polymer concentrations $c < c_0$ this limit is never relevant. Above $c = c_0$, it becomes relevant. When the elastic limit reaches point A the polymer concentration just corresponds to coils in contact $(c = c^*)$.

c_0 will be very small. But, conceptually, the existence of c_0 reveals a significant difference with the Lumley model.

If we increase c beyond c_0, the elastic limit shifts upward in Fig. 4.6, and we again find a buffer layer: we conjecture that, in this regime, dissipation is reduced.

At a certain higher concentration, the intersection of the elastic limit and the Kolmogorov limit reaches point A. At this moment, the largest eddies cease to satisfy the time criterion. We expect that any addition of polymer beyond this point will be less effective.

It turns out that the polymer concentration associated with point A is a familiar object of solution theory: it is the concentration.

$$c^* \simeq N/R^3 = a^{-3}N^{-3/4} \tag{3.6}$$

at which neighboring coils begin to overlap. Thus, in the interval $c_0 < c < c^*$ we expect a significant drag reduction, increasing steadily with c. At concentrations beyond c^*, the behavior is more complex, because some eddies are operating in a Newtonian regime, and the trivial increase of Newtonian viscosity due to the polymer may become the leading feature.

4.4 Conclusions and perspectives

4.4.1 *Elasticity versus viscosity*

The main idea of this chapter is that flexible coils, even in the dilute regime, behave elastically at high frequencies: a description of small eddies in terms of a renormalized (real) viscosity is not entirely adequate.

A Kolmogorov cascade remains unaltered by polymer additives only down to a certain limit r^{**} where the polymer stresses balance the Reynolds stresses. In the second scenario, this occurs at full stretching. Chemical degradation is probably severe in this last case.

The fate of the turbulent energy below the limiting scale r^{**} is unclear: in the first scenario, it might result from a delicate balance between elastic shock waves and rarefaction waves. In the second scenario, viscous effects may be immediately dominant.

4.4.2   *Other elastic systems*

(a) Apart from linear polymers, many other systems may show an elastic behavior at high frequencies. One obvious example is *branched* polymers, which may show an improved resistance to degradation. However, the regime of extreme deformation (the analog of the second scenario) is reached at much small elongations λ: we need a special discussion of branched systems, and the details of the branching statistics (e.g. the number of internal loops) will be important.

(b) Another family of interesting systems is obtained with *binary fluid mixtures* near a consolute point. Ruiz and Nelson (1981) have studied situations where, at the starting point, the two fluids are not fully mixed. They pointed out that the resulting concentration gradients induced elastic stresses in the system, and that these stresses *react on the turbulent field*.

(c) We may also consider a modified version of the Nelson–Ruiz problem, where the binary mixture is macroscopically homogeneous; there remains an effect due to fluctuations of concentration, with a characteristic size ξ (the correlation length). These fluctuations are remarkably similar to polymer coils: ξ is the analog of the unperturbed polymer size R. There is a characteristic time (first introduced long ago by Ferrell and Kawazaki) which is the exact analog of the Zimm time (equation (2)) with $R \rightarrow \xi$. There are elastic tensions (although, to our knowledge, the analog of the nonlinear Pincus formula, equation (16), has not been worked out). The fluctuations span all space: this is the analog of a polymer system at the concentration of first overlap $c^* = N/R^3$. In practice, the main limitation is that, to reach long times T_z (needed to satisfy the time criterion) we need a large ξ, i.e. a good temperature stabilization, in the one phase region, near the critical point.

(d) Similar effects may exist in the two-phase region, where small droplets are constantly broken (and coalesce) in turbulent flow: here, the elasticity of the droplets can be expressed in terms of one interfacial tension, following the classic papers of Taylor. Very near to the critical point, however, we should return to a microscopic description: the basic formulas are described by Aronovitz and Nelson (1984).

4.4.3 The third scenario

All our discussions have used, as a starting point, the 'time criterion' of Lumley: spatial effects have been ignored because the coil size R is usually much smaller than the smallest eddy size. However, after stretching, the situation may be different. A deformed coil of length λR, with $\lambda \gg 1$, may become comparable to the eddy size r: this may lead to a third scenario, an idea first suggested to us by E. Siggia. It is easy to construct the scale r_3^{**} at which we would meet this effect. It is far more difficult to perceive what would happen at even smaller scales. In practice, the third scenario should occur only if we achieve turbulence even at very small scales, with extremely large Reynolds numbers; it will probably be associated with strong degradation.

4.4.4 Open problems

All our discussions are extremely conjectural. (a) The very existence of a single exponent n characterizing the elongation at different scales (equation (12)) is unproven. (Among other things, we might need a family of exponents to give separate scaling laws for the various moments λ^m.) (b) The behavior of the cascade beyond the elastic threshold is entirely unclear. (c) The intermittency features which are omitted in the Kolmogorov description of the cascade may be much more significant for polymer solutions than for pure liquids.

However, we feel that the proposed scheme, the classification of scenarios, and even the tentative scaling laws which we propose, should be of some use to guide future experiments.

Apart from the elastic systems discussed here, there also exist some interesting questions with rigid rods: they do have an orientational entropy, which leads to some analog of an elastic energy in aligning flows: but they cannot follow the local deformation affinely, as done by the coils. Viscous dissipation is thus much stronger. It may be that the Lumley scheme holds for rods.

References

Chapter 1

Champetier, G. and Monnerie, L. (1969) *Introduction à la chimie macromoléculaire*, Masson, Paris.

Faraday Transactions (1979), Discussions, n. 68.

Flory, P.J. (1971) *Principles of polymer chemistry*, Cornell University Press, Ithaca.

de Gennes, P.G. (1971) *J. Chem. Phys.* **55**, 55.

de Gennes, P.G. (1977) *Macromolécules*, **9**, 587.

de Gennes, P.G. (1976–1977) course, Collège de France.

de Gennes, P.G. (1982) in J.M. Georges (ed.) *Microscopic aspects of adhesion*, Elsevier, p. 355.

de Gennes, P.G. (1984) Scaling *Concepts in Polymer Physics*, Cornell University Press, 2nd edition.

Klein, J. (1979) *Faraday Transactions*, Discussions, n. 68, J.D. Hoffman, E. Di Marzio and C. Guttman.

Chapter 2

Edsall, *et al.* (1966) *Biopolymers* **4**, (1966) 121.

Flory, P.J. (1961) *J. Polymer Sci.* **69**, 105, especially section 4.

Liquori, A.M. (1968) in A.D. Ketley (ed.) *The stereochemistry of macromolecules*, Dekker, New York, vol. 3, p. 287.

Monod, J. (1969) course, Collège de France.

Ramachandran, G.N. (1968) review in Rich and Davidson (eds.) *Structural chemistry and molecular biology*, Freeman, London, p. 77.

Stryer, L. (1968) *Ann. Biochem.* **37**, 25.

Chapter 3

Arslamov, V., Ivanova, T. and Ojar, V. (1971) *Kokl. Phys. Chem.* **198**, 502.

Brochard, F. and de Gennes, P.G. (1984) *J. Physique Lettres* **45**, L-597.

Burton, R., Folkes, M., Narb, K. and Keller, A. (1983) *J. Mater. Sci.* **18**, 315.

Cooper, W. and Nuttal, W. (1915) *J. Agr. Sci.* **7**, 219.

Derjagin, B. (1955) *J. Colloid* (USSR) **17**, 191.

Galt, J. and Maxwell, B. (1964) *Modern Plastics*, Dec. 1984.

de Gennes, P.G. (1979) *C.R. Acad. Sci. (Paris)* **288B**, 219.

de Gennes, P.G. (1984) *C. R. Acad. Sci. (Paris)* **298** (2), 39.

de Gennes, P.G. (1985a) *Rev. Mod. Phys.* **57**, 827.

de Gennes, P.G. (1985b) *C.R. Acad. Sci. (Paris)* **300**(2), 129.

Ghiradella, H., Radigan, W. and Fritsch, H. (1978) *J. Colloid Interf. Sci.* **51**, 522; this paper contains a list of earlier references by the same group.

Hervet, H. and de Gennes, P.G. (1984) *C.R. Acad. Sci. (Paris)* **299**(2), 499.

Huh, C. and Scriven, L. (1971). *Colloid Interf. Sci.* **35**, 85.

Joanny, J.F. (1985) Thesis, Paris.

Joanny, J.F. and de Gennes, P.G. (1984) *C.R. Acad. Sci. (Paris)* **299** (2), 279.

Marmur, A. (1983) *Adv. Colloid Interf. Sci.* **19**, 75.

Pashley, R. (1980) *J. Colloid Interf. Sci.* **78**, 246.

Sawicki, C. (1978) in J. Padday (ed.) *Wetting, spreading and adhesion*, Academic Press, New York.

Zisman, W. *et al.* (1964) *Adv. Chem. Ser.* **43**, 1.

Chapter 4

Ambari, A. and Guyon, E. (1982) *Ann. NY Acad. Sci.* **404**, 87.

Aronovitz, J. and Nelson, D. (1981) *Phys. Rev.* **A29**, 2012.

Bewersdorff, H.W. (1982) *Rheol. Acta* **21**, 587.

Bewersdorff, H.W. (1984) *Rheol. Acta* **23**, 183.

Daoudi, S. and Brochard, F. (1978) *Macromolecules* **11**, 751.

Flory, P.J. (1971) *Principles of polymer chemistry*, Cornell University Press, Ithaca.

de Gennes, P.G. (1974) *J. Chem. Phys.* **60**, 5030.

de Gennes, P.G. (1979) *J. Chimie Phys.* **76**, 763.

de Gennes, P.G. (1984) *Scaling concepts in polymer physics*, Cornell University Press, Ithaca, 2nd edition.

Graessley, W. (1974) *Adv. Polymer Sci.* **16**, 1.

Hinch, J. (1977) *Phys. Fluids* **20**, S22.

Lumley, J. (1969) *Ann. Rev. Fluid Mech.*, 367.

Lumley, J. (1973) *J. Polymer Sci.* **7**, 263.

McComb, W. and Rabie, L. (1979) *Phys. Fluids* (USA) **22**, 183.

Odell, J. and Keller, A. (1986) in Y. Rabin (ed.) *Polymer-flow interactions*, AIP Press, New York, p. 33.

Pincus, P. (1976) *Macromolecules* **9**, 386.

Pope, D. and Keller, A. (1978) *Coll. Pol. Sci.* **256**, 751.

Ruiz, R. and Nelson, D. (1981) *Phys. Rev.* **A24**, 2727.

Scharf, R. (1985) *Rheol. Acta* **24**, 272.

Stockmayer, W. (1976) in R. Balian and G. Weill (eds.) *Fluides Moléculaires*, Grodon and Breach, New York.

Tabor, M. and de Gennes, P.G. (1986) *Europhys. Lett.* **2**, 519.

Tennekes, H. and Lumley, J. (1972) *A first course in turbulence*, MIT Press, Cambridge, Mass.

Index